ONE MAN'S GARDEN
RAILWAYS

By the same author

Technology and the Big House in Ireland
c.1800 – c.1930

The Making of Tubby Dash

Adventures in Technology

Technology in Love

ONE MAN'S GARDEN RAILWAYS

Charles Carson

ISBN-13 987-1511469968
ISBN-10 151146996X

Published by Carricknadarriff Press

First Printed March 2015

2 4 6 8 10 9 7 5 3

charlescarson.wordpress.com

In memory of my father

Acknowledgements

With grateful thanks to my assiduous editors Caroline Knox and Alan Carson and to all those who have helped me on life's track over many years.

Contents

Burney's Bog Railway

It was the copy of Model Engineer magazine that I picked up from the news-stand in Euston Station, London, in 1962 that started my life-long interest in garden railways. Married in 1964, my wife, Ann, generously allowed the expenditure of meagre funds on a new Myford ML7 lathe and an 8' x 6' garden shed to put it in. The shed was erected in the back garden of our first home in Lestannon Avenue, Whitehead, Co. Antrim with mains electricity fed by an extension cable from the house. It was clad with cedar weatherboarding that gave off a pleasant aroma in the heat of the summer sun. The lathe cost £83.10s from A. J. Reeves & Co. of Birmingham and was paid for in three instalments. That same lathe is still in use half a century and several railways later.

My late father was then a skilled jig-and-tool fitter employed by Short Brothers & Harland, aircraft manufacturers. He took an immediate interest in my new acquisition. Soon he was turning parts for the construction of a 5" gauge locomotive. It was to be powered by a two-stroke petrol engine obtained from a power assisted bicycle. See Plate 1. The crankshaft drove two dynamos via vee belts. One was a 12 volt car dynamo for traction power and the other was a 6 volt motorcycle dynamo for the field excitation of the former. Forced air-cooling was provided by an electric motor driven fan in a sheet metal housing enclosing the cylinder. Current from the dynamo was fed to a motor mounted longitudinally in the chassis and drove two four-wheeled bogies via shafts with universal joints. The bogie axles were coupled together by chains.

Plate 1. 'Diesel-electric' chassis

Driving that locomotive was a delight. First, the engine throttle was opened and then the field-current rheostat was advanced. This was accompanied by the sound of the revolutions reducing as the dynamo loaded the engine and the train smoothly took off. A superb stainless-steel body riveted together and complete with windows, lights and two-tone horn completed the locomotive. See Plate 11. My father passed away in 1981, but his 'diesel-electric' locomotive still runs on the raised track of the Model Engineering Society of Northern Ireland (MESNI) at their Cultra venue.

By January 1968 we had moved our growing family to Carolhill Road, Glengormely, Co, Antrim. This was a new semi-detached bungalow on a development built on the site of Burney's Bog, arguably one of the coldest places in Ireland in the wintertime. Proof that it was indeed an ancient bog appeared as large pieces of black bog-oak were found on the undeveloped ground in front of our home. The workshop shed was re-erected on a concrete foundation provided, for a small payment, by the development's builders who were still on site. Later we aspired to a brick-built garage, greatly increasing workshop and domestic storage space. It never accommodated our car.

Soon thoughts turned to building my first railway. No streamlined record-breaking trains for me however. Ride-on rackety narrow-gauge with sharp radius curves was my thing. Above all I wanted to see the world from the perspective of the driver with eyes three feet above the track. A track that guided one through a time-tunnel into another dimension. A phenomenon that actually seemed to happen, especially at night, with a dim headlamp and a couple of pints of home-brew consumed.

It has often been remarked that we garden railway enthusiasts belong to a very broad church. I found myself a bit to the east of model engineering and far to the south of 'N' gauge. There was little enthusiasm for gardening, since it always appeared that preparations needed to have been made months prior to planting. Although still wanting to run in the great outdoors, there were never going to be miniature forests or sculptured upland lawns. Unable to afford ready-made ride-on locomotives and track, it was to be a do-it-yourself operation. The focus was to be on minimum cost with maximum enjoyment. Although appearance would be important, there would be no 'rivet counting'. So, even though most model engineers of fifty years ago required 5" gauge railways to be elevated for stability reasons, the decision was taken to settle for a 5" gauge ground-level track. This meant an oval of 30' x 20' could be accommodated in the back garden with a siding into the new detached garage. See Plate 2.

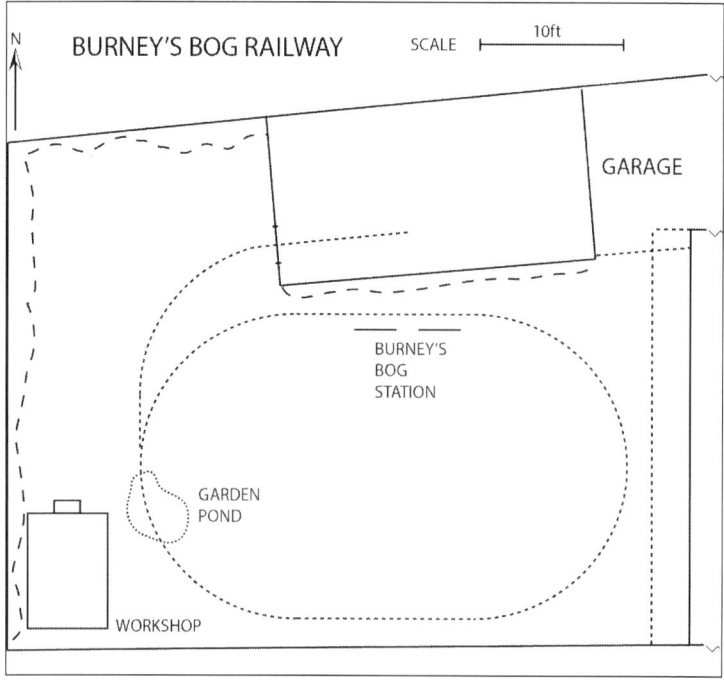

Plate 2. Burney's Bog Railway

The track was made from black iron 1" x ¼" flats and ⅜" diameter spacers with turned down ends riveted in place. A very rigid track resulted, which greatly reduced civil engineering work on the uneven surface of the garden. Very few supports were needed.

Some people buy a ready-to-run or a kit locomotive out of a catalogue. Others purchase plans and castings to make one for themselves. However, the 'odd' person simply consults the scrap box and a few magazine articles and builds 'out of their head'. It so happened that there was an ex-government surplus aircraft flap-motor and a dubious 12 volt car battery on the garage shelf. So that was OK. All that was needed was the rest!

The general appearance was simply determined as it was going to be an electric 0-4-0 locomotive with the battery under a bonnet and the driver's cab behind. Anticipating many years of arduous service on imaginary narrow gauge lines in Ireland's gardens, it was considered important to have a 'proper' chassis. Normal model engineering principles were followed, in that frames were hack-sawed and filed from ¼" thick black-iron complete with openings for axle boxes. A seriously arduous task, whereas nowadays CNC water-jet cutting machines can cheaply profile any shape with practically no need for finishing.

Before proceeding further, some elementary mathematics were needed to work out the necessary gearing to achieve a reasonable speed on the track. What was required was a stately progress through the landscape of say around 4mph. Even in the days when 'health and safety' was unheard of, one wanted to avoid damage due to youthful speedsters (ahem!) derailing on curves. The motor plate stated that the speed was 1500rpm at 24V; say 750rpm at 12 volt without reduction gearing. Since the diameter of the proposed wheels was to be 3.5", the circumference therefore would be 3.142 x 3.5" equalling near enough 11". At 750rpm the wheel would be traversing 750 x 11" = 8250" of track per minute. This equates to 7.81mph. To achieve the desired maximum speed of 4mph therefore, a reduction ratio of 2:1 was required.

Four wheels were turned from 3.5" diameter steel bar and fitted to the axles. This 'fitting' was a nerve-wracking business in those days before 'metal glue' such as 'Loctite'. Having achieved axle and wheel bore dimensions alleged to give an interference fit, the wheel was heated by use of a propane torch and the axle cooled in the fridge. Then the two were 'assembled'. Nowadays one hears much of 'white coat syndrome' affecting blood pressure readings in the doctor's surgery, but little of 'shop coat syndrome' afflicting those attempting to 'assemble' wheels and axles in the home workshop. Unless all the parameters are correct; interference fit, hot enough wheel, cold enough axle, speed of bringing them together and a sober mechanic equipped with leather gloves (although not so sober that he or she fails to attempt the operation) the following results may be obtained. The axle completely refuses to enter the wheel-bore even when persuasion is attempted with a heavy mallet. Next to complete and immediate success this is probably the next best result. The recommended procedure to follow in this case is a beverage of your choice, followed by an early night and a retry the following day. With the offending parameter adjusted, victory should be achieved. Highly undesirable is the situation when the axle half-enters the wheel and totally refuses to advance a millimetre (or a few 'thou.' in 'old money') regardless of violent persuasion by the heavy mallet aided by enraged vocal encouragement. With wheel sets complete it was back to the main locomotive construction.

The reduction ratio was to be achieved by worm and pinion gearing, the worm gear mounted on the motor shaft. One disadvantage of this arrangement was to become obvious later. When the power to the motor was cut the train could not coast under its own momentum due to the pinion wheel being unable to drive the worm wheel. On the other hand, power reduction produced an effective braking action. A controller handle managed direction and speed of the engine. This swung a phosphor-bronze wiper over stud contacts made from 4BA cheese-head brass screws and nuts. These were arranged in an arc through holes in a 'Paxolin' panel. Movement of the handle introduced or cut out high wattage series resistors, home-made from old storage heater elements, connected between the studs.

The locomotive sported cast-bronze 'name plates' on either side of the bonnet, which read BBC. See Plate 3 where 'scale driver' Caroline is at the controls. This quaint device resulted from finding the plates from an old discarded studio-microphone at the BBC where I was on the staff as an engineer at that time. The cab was made of wood with a curved sheet-steel roof and painted sky blue. Spectacle frames were lid-rims from cocoa tins with the centres removed. A complementary yellow was used for the chassis, dark

Plate 3. 'BBC' on garage siding

Plate 4. Burney's Bog Station

blue for the bonnet and dark red for the axle boxes.
The choice of colours was dictated not so much by

artistic design but more by the remnants of domestic paints on the garage shelf. The front grille was made from expanded steel sheet and the working headlamp a bullseye lens mounted in a shoe polish tin.

No garden railway is complete without a station as a focus for train ride activities. This one evolved as seen in Plate 4 where 'BBC' is halted in Burney's Bog Station with pretty Gillian Jennings perched on the driving truck. Eldest son Andrew is the ticket collector. Note the hand-painted nameplate illuminated by a low voltage bulb in an upside down jam jar mounted on a bracket on top of the post. Station fencing, made from pointed wooden laths, seems to be supported by the uprights being jammed in the holes in left over building bricks. By the year of this photograph, 1973, the contents of the wooden workshop had been moved into the back of the garage. My ML7 lathe is just visible in the left hand window.

Unseen in the picture, but hiding below the right hand window, is a 'Brasted' iron-framed upright piano. If one imagines setting off from the station, the track curved to the left over the garage siding point and then traversed a bridge over a garden pond. In the dusk of one memorable summer's evening, consumption of home-brewed beer combined with enthusiastic playing from the garage piano, led to passengers aboard a night-special train being deposited in the pool. The 'Brasted' piano had an eventful journey ahead of it before it reached its final resting place high on a drumlin in Co. Down.

Occasionally, 'BBC' appeared at the raised track of the MESNI at Cultra. This attractive venue was, and is still, inside the walled garden of a demolished big house called Dalchoolin, which reputedly was the childhood home of my great grandmother. See Plate 5.

Plate 5. Dalchoolin 1971

Plate 6. BBC at Cultra 1972

Mind you, there were also family rumours of 'skeletons in the cupboard' and an alleged elopement with the gardener. The site is part of the Folk and

Transport Museum complex. Plate 6 shows 'BBC' on this track in the walled garden during Whitsun 1972. On board, behind Andrew at the throttle are myself and my mother.

The years passed quickly in Glengormely and with an increasing family our 750 square-foot bungalow began to seem a little cramped for space. In January 1974 we moved to a new life on a windy hill off Ballycrune Road, Carricknadarriff, County Down. Suffice to say that building railways and model engineering activities were out of necessity curtailed as my wife and I tackled the transformation of a derelict farmhouse into a home. However, one attempt was made to make it easier to keep running at the MESNI track at Cultra, namely 'BAT'.

'BAT' was my second battery-electric locomotive. See Plate 7. It was loosely based on industrial locomotives as used in mines and factories. Although not a 'thing of beauty', indeed one child onlooker at Cultra was overheard telling his father "that's not a real train", 'BAT' possessed certain advantages. The weight of a locomotive directly affects the tractive effort, but unfortunately it is also injurious to the back of someone trying to lift it. This type of locomotive used a battery unit that could be swapped with a fully charged one. When the heavy car battery was removed from the motorised chassis the units could easily be loaded in the car boot (a Mini) for the journey to Cultra. Instead of axle-boxes sliding in slots in the frames, the wheel bearings were mounted at the ends of swinging arms with rubber blocks providing springing. In other ways too it was a considerable advance on 'BBC' in that it had a home-constructed electronic speed controller, an ammeter and a battery-condition meter. The story of the fraught making of the controller, in particular the etched 'printed' circuit board was told in a series of articles entitled 'Bats in the Belfry'. These were published in a series of three parts in The Link, the journal of the MESNI.

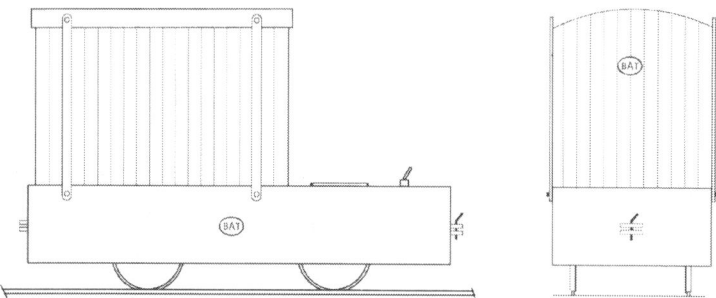

Plate 7. Author's drawing of 'BAT'

Windy Hill Railway

Six years were to pass before it was possible to resume 'normal' activities. After years of operating battery-electric powered locomotives that had the advantage of instant readiness, provided the battery was charged, a great desire to build a steam locomotive arose. Something that lived and breathed!

Beginning in 1980, several years and many, many man-hours were spent in building a 5" narrow gauge 0-4-0 saddle tank locomotive. This was a very successful design called 'Sweet Pea', marketed by Blackgates Engineering. Such was the enthusiasm that I often got up at 5.30am on summer mornings to continue work in the workshop. This was the third locomotive to be constructed. Sadly my father passed away at the age of sixty-six in 1981, after only one year of retirement. His help and enthusiasm had enabled me to overcome my reluctance to tackle a steam locomotive. He was an Engineer.

I called the locomotive 'Friar Tuck' because of its rather portly appearance and brown 'habit' (paint job). The finished engine was a great success, pulling heavy passenger trains at Cultra. On one occasion there were sixteen adults entrained. Later it proved a capable performer on the rather steep grades of the garden railway at Ballycrune Road, which was built on top of a drumlin. 'Friar Tuck' was loved by all, the sheer animalistic power as he pulled away with a heavy load and the shriek of his steam whistle made an indelible impression on children and adults alike. My son, Andrew, was a frequent and enthusiastic driver. He was highly competent, could be relied on to keep a 'full glass of water' and a good fire in the firebox. Some of my BBC colleagues drove him too. See Plate 8.

Plate 8. Peter, Frank and author steaming

The photo shows Peter Lindsay, at that time a BBC TV film sound recordist, at the throttle. Peter now works in the movie industry as a Production Sound Mixer, including on such Hollywood blockbusters as Avengers: Age of Ultron. The passengers riding 'beer-crate' trucks are Frank Murphy, another BBC colleague – now a retired Picture Editor - and myself enjoying a pint at the rear. The rest of this chapter is an updated reproduction of an article that appeared in Engineering in Miniature in September 1984, entitled

'Windy Hill Railway'.

Our home is situated on the top of a drumlin. What are drumlins? They are half-egg shaped hills formed by glacial action, and this part of Co. Down has been so liberally bestowed with them that there is a striking resemblance to a basket of eggs. However, one cold and blustery night in 1982, when work on 'Sweet Pea' had ceased for the night, my eldest son Andrew and I decided to construct a railway and examined the various possibilities. The workshop had a concrete floor, and outside the back door was a concrete path. This was bordered by an orchard from which the apple trees had been cleared, but had become so overgrown that it had defeated all efforts in the past to tame it. We decided to use the railway not only to gain entry to this ground, but also to enable us to move the materials to create some order out of chaos. The gauge was decided for us by 'Sweet Pea', and the inheritance of a magnificent 'diesel-electric' locomotive from my late father. So the idea of The Windy Hill Railway (WHR) was born. A 5" narrow gauge contractor's line to be built in three stages, as simply and inexpensively as possible, as befitted the pockets and navvying abilities of one man and a boy. See Plate 9.

The first stage was a straight tramway across the floor of the workshop and through the door, coming to a stop at the end of the concrete path ready for a possible future extension in stage three. A visit to a local steel stockist produced the ½" x $3/16$" black mild steel to be laid on its flat for the tramway section, and although I was doubtful I agreed to purchase a large quantity of 1" x $^3/_{16}$" material for the rails, when informed that "the next lot in would be 50% more expensive".

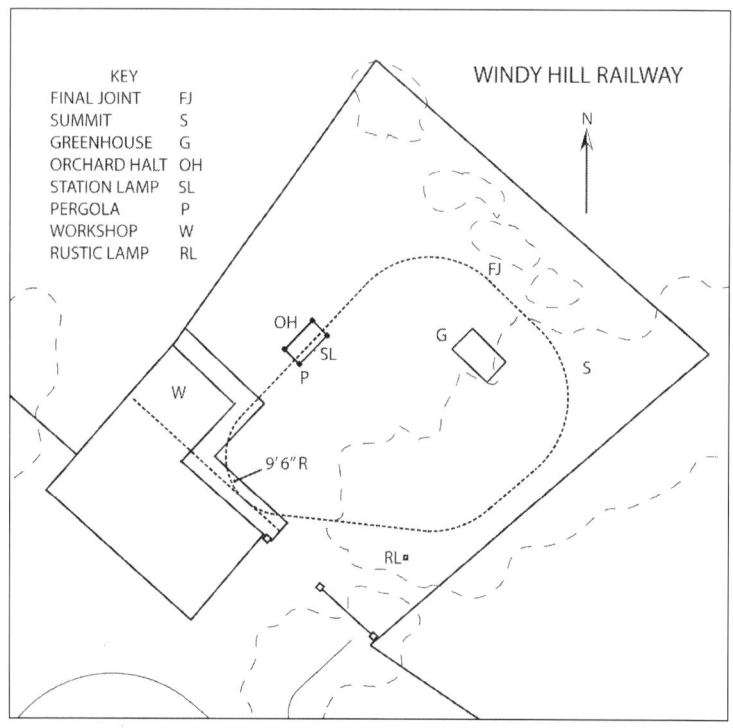

Plate 9. Windy Hill Railway

Stage one was rapidly accomplished with Andrew using a portable electric drill on a stand, drilling No. 8 screw holes complete with countersinks at 12 inch centres, while I drilled and plugged the concrete. Some initial trouble with the masonry bit wandering in the concrete was cured by standing on a steel plate in which a suitable hole had been drilled to give the necessary guidance. 'Sweet Pea' was by the summer of

1982 a rolling chassis, and we now had a short track on which to try it.

Plate 10. 'Friar Tuck' approaching the 'flat points'

After a little trial and error, we succeeded in making two pairs of hand operated points which, apart from the weighted levers, were as flat as rest of the track that is $^3/_{16}$" proud of the concrete. See Plate 10. These have given no trouble to foot traffic at all and the only maintenance required has been the occasional 'blowout' with the workshop air line and a drop of oil. These points were 'built on the job' to a nine-foot six inch radius, which is the minimum on the line, and the other curves to a fifteen-foot radius. We were now ready to strike out across country on stage two.

I have mentioned doubt about the 1" x $^3/_{16}$" black iron flat. This was for two reasons. First the 1" dimension seemed rather high from an appearance point of view, and secondly the $^3/_{16}$" seemed somewhat narrow for good tractive effort. Both fears have proved groundless. The height, in fact, was an advantage with the method of track construction chosen and on measurement using a spring balance with the 'diesel-electric' locomotive no reduction of draw-bar pull could be detected. Some years previously a large quantity of random off-cuts of native timber was obtained as firewood from a local sawmill. This timber was to prove useful to railway builders in many unexpected ways.

The worst part of the whole undertaking was undoubtedly the track bed. After marking out the route with pegs and a long lath as a giant compass, it was found that the maximum gradient was 1 in 70, which was satisfactory. Several attempts to penetrate the matted layer of grass were discouraging, and resort was made to a garden fork to mark a two-foot wide strip of 'turf' to be removed. The garden spade was taken to the workshop grinder and sharpened. Using the spade in the manner of a huge wood chisel it was found possible to 'shave' a 4" deep roll of soil and couch grass from the track bed.

Since considerable night running was contemplated, this was the stage to install the low voltage under track cable to supply various lighting points. Standard 2.5mm twin-and-earth mains power cable was laid from the workshop direct on the soil and covered with flat stones for protection. At six points, loops were left for connections and then the hard core was filled in to within 2" of sleeper level. Using this cable fed from a battery charger across an old car battery, it was possible to provide two independent 12 volt circuits around the track. One switched from the workshop for the track lighting; the other to supply a feed to local switches in the greenhouse, etc.

After much consideration the final method of track construction, pressing the rails into pre-prepared slots in the sleepers, was decided. It was to prove simple, inexpensive and trouble free. A wooden jig was made that would admit timber (firewood) up to ⅞" thick, and any lengths which would enter the jig and were of a reasonable width, approximately 1¾", were set aside as suitable for sleepers. See Plate 11. To use the jig, timber was fed in to the end stop. Using a wooden lever, Andrew ensured that the embryo sleeper was held flat on the bottom and hard against the side guide. A portable circular saw, fitted with wobble washers then made two cuts against the side guides producing the bare $3^3/_{16}$" grooves at $5^3/_{16}$" centres. A hand saw, in another guide, cut off the 13" length and the finished sleeper was pushed out. Two three-hour sessions made all the sleepers that were required and it only remained to place them in the creosote bath for several days, followed by draining and drying, before the track construction could begin.

The advantage of this rather crude jig was its ability to accommodate various thicknesses and widths of timber. Provided there was a flat bottom face it did not matter if the top was a bit curved or one edge was not parallel. We actually selected rustic looking pieces and the final result had that narrow gauge look we were seeking.

Plate 11. Sleeper jig

Using a simple lever press welded up from scrap it was possible to construct the track in situ at a surprising rate. At first we tried to make the curves by using concrete blocks to hold the rails in position whilst the sleepers were fitted. It was soon found to be simpler to make the track straight and, by gently tapping with a plastic mallet, the curves could be produced accurately to the required radius. As each length of track was finished a truck carrying the electric welder was brought up and the ends joined, until the track loop was complete.

Plate 12. Orchard Halt

Plate 13. View from summit of line

After the 'diesel-electric' locomotive had broken a paper tape across the track, a dated brass plate was screwed to the sleeper nearest to the final joint and stage two was complete. See Plates 9, 12 and 13. The timber mentioned earlier was now used to construct a pergola over the track and a name board 'Orchard Halt' was fitted, illuminated by a scale sheet brass and copper lantern. Several fences were added to give the track more scenic interest and a rustic wooden station lantern installed to light a trackside barbecue area. The total cost of the project was £25 (about £75 in 2014) plus a tremendous amount of very hard work.

Changing Times

The years passed. In 1979, at the age of forty-one, I had resigned my BBC staff job as an electronics engineer. I bought a kit of lights and took up freelance film lighting. This is not the place to tell of the exciting and often dangerous situations that were encountered working with local and foreign film crews in the 1980s. But all things come to an end. The new ENG (Electronic News Gathering) cameras could produce acceptable broadcast quality TV pictures with available light. Even for documentary programmes the cameraman now carried his own lighting kit; the 'lighting man' was no longer required.

Architectural metal work, specially made fittings and quality wrought iron work, proved successful enough as a replacement for the lighting, but it was difficult to keep a continuity of good paying commissions. Three examples can be seen in Plate 14. To the left is one of a pair of brass luminaires made for the Parish Church in Annahilt, County Down. In the centre is a brass weather vane commissioned for the Balmoral Golf Club, Belfast. To the right is one of several brass luminaires with a maple-leaf motif made for the Canada Room, Queen's University Belfast. These and many other similar items are still visible to this day.

Eventually, 'Friar Tuck' was sold at auction at

To view a video showing how the author made his weather vanes - visit:
https://youtu.be/7WFr91P4Sgs?t=20s

Christies in England for £1000. Thus realising about ten pence per hour for all that work!

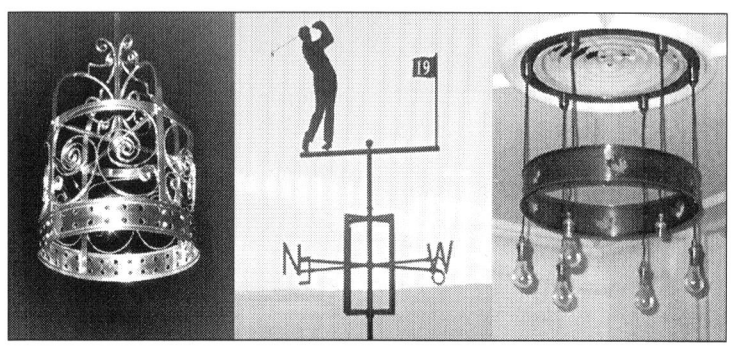

Plate 14. Architectural metal work

By the mid-80s the WHR track had been invaded by a tough carpet of couch grass and in places the track was invisible. One summer's day however, what started as a picnic in the wilds of the orchard turned into a railway day. Willing hands tore at the grass and gradually exposed a rusty but intact track. True, many of the sleepers had rotted but were adjudged to be just sound enough for running. My Dad's 'diesel-electric' was fired up and a dusty driving truck was produced. Sure-footed as always, the loco negotiated circuits of the track giving great pleasure to all. Sadly however, in the days that followed it was decided that the track was too far gone to restore. Perhaps this was the time to make a change to the larger and more stable 7¼" gauge?

Carricknadarriff Tramway

Good times returned when I started working in the BBC as a contractor providing broadcast engineering services in the early 1990s. A renewed love affair with garden railways began, but this time in 7¼" gauge. Since the 'diesel-electric', being 5" gauge, would have nowhere to run at home and it would have been difficult to transport, it was sold to a fellow member of the MESNI who would run it at Cultra. Thus I would continue to see it operate and happily still do thirty years later.

The first venture in the larger gauge, in the mid-1990s, was a battery-electric tram with a roof. It carried the number '3', being the third electric vehicle to be built. 'BBC' and 'BAT' were innocent of numbers and 'Friar Tuck' carried No. 1 as the first steam locomotive. It had never been anticipated that the railway building folly was to be for life! No. 3 tram was propelled by a converted 24 volt dynamo that I had amongst the bits-and-pieces that had been inherited from my father. It was powered by a 12 volt car battery. A drum controller was made as being more prototypical than an electronic one, but it did not offer the smooth control of the latter. However, it was not prone to expire, issuing expensive bad smells and smoke, as the early home-built electronics ones occasionally did.

Unfortunately this vehicle suffered from a couple of design problems. Chains and sprockets were used for a two-stage reduction drive from the motor to the wheels. There was a large sprocket on one of the axles and ground clearance was minimal, especially when there was vertical movement due to the springing on track undulations. A second difficulty was to keep the tensions correct on the various chains as wear took place. Adjustment of one chain affected that on the next stage and so on. Two slatted cross-bench seats accommodated the driver and one adult or two small children. The curved end panels were made from sheet steel and were each supported by six ⅜" diameter vertical rods topped by varnished mahogany bows. A brass No. 3 house-door number and a lamp made from rolled up brass sheet were mounted on both panels. Braking is applied to all four wheels by a purposeful looking brass tram-handle via a pawl-and-ratchet wheel that was filched from a wire-strainer in the sheep fence at the garden gate. A simple white-painted hardboard roof curved over ¾" softwood frame members was carried on four supports. These are made from an upper ¾" diameter steel tube welded into a 1" square steel tube bottom section. At the roof end there is simple decorative scroll work.

Since there was as yet no home railway to conduct trials, No. 3 was transported by car and trailer to the track at Cultra where a few adjustments seemed to cure the teething troubles. See Plate 15.

Plate 15. Tram No. 3 at Cultra

Early in the life of the tramway, about 1997, it became obvious that No. 3 was going to be rather lacking in power to deal with the incline and the fifteen-foot radius curves. So it was decided to offer it for sale to raise funds for a replacement. It so happened that a friend of mine, the chairman of a well known local 7¼" gauge miniature railway, knew of a young chap who was not a model engineer but was 'mad keen to get on the rails'. One evening, shortly before Christmas, the chairman appeared with a car and a trailer. He purchased tram No. 3 and landed it in the young man's driveway at dead of night so that it would be discovered the next morning. Santa's toys had just got bigger.

It was decided to build a replacement for No. 3 that would be ready to run when the new track was completed. Thus it was that No. 4 tram, as yet unpainted, got its first outing during a 1999 mid-summer party. About forty feet of temporary track was laid in the orchard to give a short there-and-back run. A rustic outdoor bar was constructed and decorated with tree fronds. It had candle lanterns ready to be lit when darkness fell. The guests, including Andrew's boss and his wife who were visiting from San Diego in the USA, were due to arrive around 8.00pm. Ominous clouds gathered overhead as the arrival time approached. With perfect timing, heavy rain started to fall as the guests arrived and continued until nearly midnight. It was a 'wash-out'.

The only redeeming feature was that 'Mrs Boss', probably feeling sorry for us, braved a trip with a tarpaulin draped over one side of the tram to keep out most of the driving rain. Of course, the sun came out 'a blinder' next morning and shone all day long. Tram No. 4 was in a sorry state. The unpainted steelwork had rusted overnight and stained any timber with which it was in contact. It took days of work to dry everything out and mitigate the damage.

Perhaps the 'highlight' of the evening was the competition held in the workshop, rain hammering down on the corrugated-iron roof, to see who could sign their name with my 150 amp MIG welder on a steel sheet. Hardly the imagined balmy summer's evening!

The new 7¼" gauge track starting point was the same as the previous 5" gauge railway, in the workshop. See Plate 16.

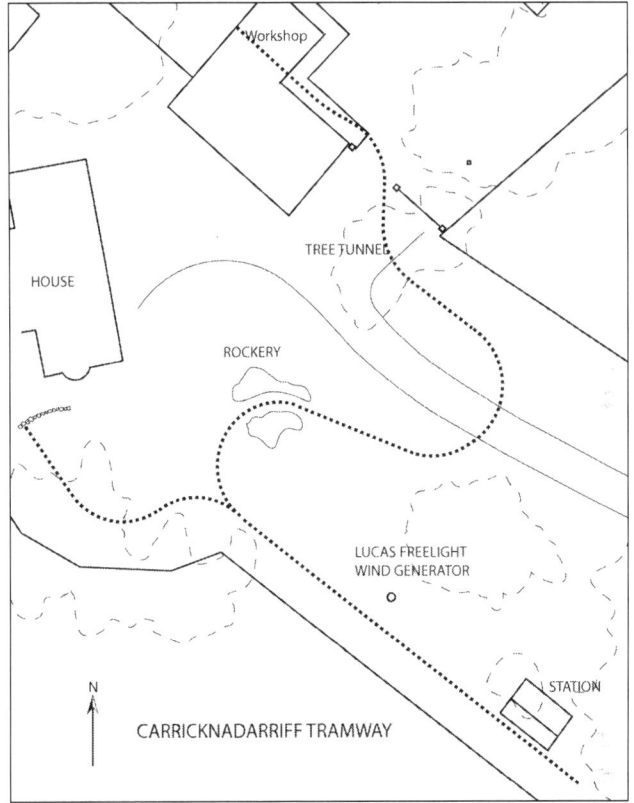

Workshop

TREE TUNNEL

HOUSE

ROCKERY

LUCAS FREELIGHT
WIND GENERATOR

N

STATION

CARRICKNADARRIFF TRAMWAY

Plate 16. Carricknadarriff Tramway

This time however, the track was made up by MIG welding 25mm x 6mm rails to 25mm x 3mm sleepers all supplied as black mild steel flat. Because it was the workshop door, thus not of 'domestic authority' concern, the door was shortened to allow it to close over the track. A counter-balanced flap with slots in it was arranged to drop down to complete the closure and deny entry to mice. This welded construction was carried out 'on the job' to suit the levels and curves required. Mostly this meant kneeling on the ground as it was in the winter. A four wheel truck carried the MIG welder to the 'head of steel'.

Curves were formed by clamping on an extra length of rail, pulling the required radius, and holding all in place with multiple concrete blocks whilst welding. Each track panel had the first and last sleepers affixed right at the ends of the rails. Short pieces of ¾" diameter steel bar with a ⅜" hole drilled in the lathe were prepared. These were welded longitudinally to the sleeper ends enabling 8mm stainless steel bolts to clamp the panel's ends together

in alignment.

Plate 17. Tree Tunnel

It was a pleasure later to ride on the track in

glorious summer sunshine with only distant memories of drifting snowflakes and freezing hands and knees during the construction. This type of track was well suited to crossing the gravelled parking area and the compacted stone driveway as it could withstand being constantly driven over by cars and delivery lorries. A local small engineering firm obligingly arranged for my track panels to be galvanised along with some of their items at a very modest cost.

As can be seen in Plate 16 and Plate 17, the track passes through a 'tree tunnel'. This proved to be one of the best features of the line. The approach was usually in deep shadow from the trees, and then the train travelled through the stygian gloom of the tunnel and suddenly burst out into a sun-soaked open landscape. An exotic variant could be enjoyed on nights of the full moon. These delights were only available of course when the sky was not overcast with eight octa cloud at one hundred feet or when it was not raining, snowing or foggy.

At the end of the right hand curve after crossing the driveway, the track construction changed to ⅝" high extruded aluminium rail screwed to 2" x 1" keruing hardwood sleepers. Fixing was by 5mm x 25mm long stainless steel pan-head screws and 6mm washers. When the property was first purchased many 'Castlewellan Gold' trees were planted. Hindsight is a wonderful thing, but at the time we were unaware of the enormous growth potential of this species. The area marked 'rockery' was one such place that had to have a felling of several sixty-foot high specimens. To avoid the expense of a contractor 'grubbing up' the roots, the stumps were cut off as close to the ground as possible and covered with builder's polythene. Large rocks were placed on top and in filled with a mixture of soil and compost. Planted out with various heathers and ground cover plants there was soon an attractive feature to drive through.

The serpentine route of the track was decided partly by the need to minimise the gradient and partly to take in the most attractive features of the garden. As it turned out, there were still times when heavily laden trains only just managed to maintain traction when the track was wet. Leaving the rockery curve behind, the track ran straight along the 'march ditch' at the south-western edge of the property, past the 'Lucas Freelight' wind generator and terminated at the station. The site of the station commanded an extensive southern view of the valley and the Mourne Mountains in the distance.

Younger son Alan and I mixed the concrete for a raft base on site. We built the small station building seven courses high with red brick recovered from house alterations, and the upper part was timber clad as in garden sheds. Two recovered mahogany exterior doors fitted side-by-side opened unto the platform and another door on the opposite side unto the garden. This project made much use of the heavy load carrying capacity of the railway. As it was envisaged that the station building would double as a miniature 'village hall' for entertainment purposes, an underground cable was laid from the garage via two outdoor 13 amp sockets, one at the rockery and one at the wind generator, terminating in the station. Each of these locations was lit with a mini lantern. The waterproof 13 amp socket boxes were mounted on large rocks that had a vertical face. For safety reasons, the garage end of the underground cable was connected to the mains electricity via a 30mA RCD.

Work on Tram No. 4 was completed in time for Christmas 2000 and it was named 'Jessica', in honour of our first grandchild. The name board on the side panel was machined in wood using a miniature CNC router that had recently been constructed. That Christmas many of my relatives received varnished mahogany letter-racks with their names picked out in white, created using the router.

Tram No. 4 incorporated several refinements to correct the design deficiencies of No. 3. There was a centre back-to-back cross-bench seat, and a more elaborate clerestory roof held aloft by eight tubular supports with brass finishers. Under the seat, was a third-horsepower 24 volt Bosch DC electric motor supplied by two 12 volt 'Leisure' batteries that greatly enhanced load carrying and hill-climbing abilities. This time, an electronic speed controller from 'Parkside Railways', was mounted on the front panel bow giving smooth acceleration in both directions.

Adjustable chain tension was provided to each axle by mounting them on separate sub-frames and to the counter-shaft by an adjustable motor mount. Ground clearance was also improved by using larger wheels. The overall result was a startling 'bat-out-of-Hell' performance. Unfortunately, this led to the first accident on the Carricknadarriff Tramway Company's track and a strange sequel to the Santa Claus story already mentioned!

One autumn evening 'the chairman' went for a trip up the line. On the way back running in reverse, having passed through the tree-tunnel and onto the sharp right-hand downgrade curve, a momentary confusion occurred on the part of the driver and full power was applied instead of braking. Tram No. 4 couped. Luckily the only serious damage was the hole in the roof where 'the chairman's' head had passed through.

Plate 18. Tram No. 4 (Jessica)

One day close to Christmas that year, a phone call established my willingness to part with the now-roofless tram. Near to midnight a further call warned of imminent collection. Uncle Bob and I, who had been having a few early festive pints, found ourselves assisting with the loading of the heavy vehicle into the chairman's car trailer. It was reported that Tram No. 4, operated the 'Santa from the North Pole' special at Drumawhey Junction the following day. I once had the opportunity to drive this tram on that magnificent mile-long track and was able to 'give her the holly'. The performance was all that could be desired. Plate 18 shows my wife, Ann, in the passenger seat of No. 4 at Drumawhey Junction, its new home.

And so the opportunity arose to construct yet another tram. But it was to be delayed while a very different size of garden railway appeared, had its days in the sun and disappeared. It all started with a toy Chinese 'Rocky Mountain' locomotive resplendent in red and green plastic. The 'pros' were that it had a very realistic air operated chime whistle and an effective smoke effect unit. The 'con' was that it was not 'G' gauge, that is 45 mm or 1.772" between the rails. This was disappointing, as I had hoped to entertain the grandchildren (ahem!) by having it run around the 2002 Christmas tree. Despite a careful study of the plastic locomotive chassis there seemed to be no way to simply reduce the gauge to suit my circle of brass-railed LGB 'G' gauge track. Leafing through a Model Engineer magazine provided a clue when I found an article describing how a run a 5" gauge locomotive on 7¼" track. Use a convertor bogie! Next question; where was such a specialised bogie to suit my conversion going to come from? Answer; make it!

The idea of turning up such small wheels on

the lathe for a bogie and then possibly many more for home made trucks did not appeal. Thoughts turned to injection moulding, a technology that I had always wanted to try. A short description of the process in a model railway magazine steered me in the right direction. Briefly, pelleted recycled plastic is put in a cylindrical brass cup silver-soldered to the bit holder of a large 125W 'Solon' soldering iron. An attached kitchen oven thermometer enables an eye to be kept on the temperature of the molten plastic. See Plate 19.

Plate 19. 125W iron with cup & thermometer

The cylindrical cup has a small nipple at the bottom. A separate close fitting piston is mounted in the chuck of an ordinary drill press. A two-part steel mould, bolted together, with the internal dimensions machined to the wheel profile, has an orifice (the 'gate') to suit the cup nipple and a small 'sprue' hole to allow air to escape during filling. The assembled mould is placed on the drill machine table. When the plastic has melted, the soldering iron/cup is held above the mould with the piston just entered in the cylinder and the nipple inserted in the gate. The drill-press lever is operated downwards until molten plastic emerges from the sprue hole. Within a short time the bolts can be undone and the finished wheel pops out. Provided the mould has been carefully made, no further finishing work is required other than to cut off a small piece of sprue. The self-coloured wheels are then force-fitted to their axles. See Plate 20.

Plate 20. Injection moulded 'G' gauge wheel

Surprisingly the 'Rocky Mountain' locomotive performed well when perched on top of its convertor. It had no problem pulling several trucks loaded with miniature Christmas goodies; pine cones, little foil-wrapped Santas and wooden toys. An unexpected result of the conversion was that the direction of motion was reversed. Thus proceeding in reverse when the intention was to go forwards, an Irish-American locomotive perhaps?

Soon after Christmas, the outdoors beckoned. 'G' gauge was really for garden railways wasn't it? So lengths of LGB track were purchased and laid on treated timbers intended for garden decking. See Plate 21.

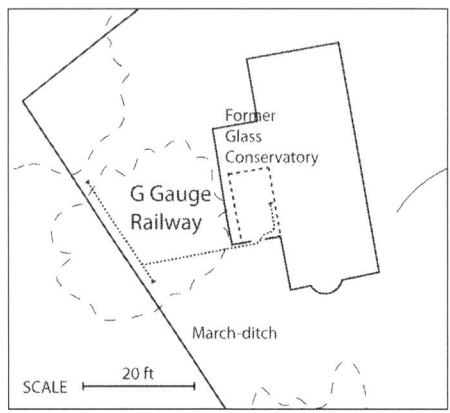

Plate 21. 'G' gauge garden railway line

A lift-out section of track enabled trains to pass through the opened window of the aluminium and glass conservatory at the back of the house. A straight run of track was laid along the top of the south-western march-ditch bank. Any miniature passengers would have had pleasant views of grazing sheep in the foreground, with Slieve Croob behind and the Mourne Mountains in the far distance.

If one could injection-mould wheels why not other items? It was decided to build a 0-4-0 'Plymouth' locomotive to suit the 'G' gauge track, with an all-brass body and as many of the other parts as possible produced by injection moulding. Steel moulds were made for the wheels, coupling rods, cranks and the multi-height industrial style coupling blocks. See Plate 22 for an example of a coupling block produced in this way.

Plate 22. Injection moulded coupling-block

'Plymouth' was equipped with a sound card that reproduced the sound of a bell triggered by a magnet block between the rails. At coupling level front and rear were horizontal pins that operated reversing switches allowing unsupervised running on the end-to-end track. The American theme was enhanced when Alan brought a model 'Aero' wind pump home from the USA.

 Take a short trip on the line - visit:
https://youtu.be/EqTKR8kM-Vc

Nice as it was to see the view on the railway as caught by the camera car, there was an urge to actually be on board oneself. That meant a return to 'the big one', ride-on 7¼" gauge. And a new and better tram! All the G gauge equipment was dismantled and sold on eBay as seen in Plate 23. All, that is, except a small circle of track, the 'Rocky Mountain" locomotive and a home made large capacity hopper car. This much reduced outfit has circled many Christmas trees ever since.

Plate 23. Sale of G gauge railway on eBay

The new tram was to have all the same refinements as No. 4, plus regenerative braking that worked in the following way; reducing power on the controller caused the motor to operate as a dynamo thus applying a braking effect to a heavily loaded train. Cleverly the roof struts were not sheeted, thus enabling high speed running without danger of head-through-the-roof embarrassment on downgrade curves. The usual house-door brass number, No. 6 in this case, was fitted front and rear. Why was there no No. 5? Possibly because the hardware shop had run out of fives that day. Construction of this mini-tram took about six months during the winter of 2004-5.

The design challenge was to create a 7¼" rail vehicle that would comfortably accommodate my 6' 1" frame and size 11 shoes with the minimum external dimensions. Since no other passenger vehicle existed at that time, a second seat was desired for a child. See Plate 24.

Plate 24. Tram No. 6 halted at the rockery

A further important requirement was that the tram should be capable of pulling one, or possibly two, trailer coaches up the steep grades of the Carricknadarriff Tramway Company's line in the future. With this in mind, a 12 volt half-horsepower EMU traction motor was installed.

Sticking to the tried and tested arrangement of Tram No. 4, the axles ran in ballraces mounted in machined cast iron axle boxes that moved vertically in horns controlled by compression springs. The horns and diagonal bracing pieces were welded up from black-steel 16 x 5mm flat bar and attached to the 30 x 5mm angle sub-frames. Each sub-frame was bolted to a main chassis with slotted holes to allow longitudinal adjustment when the bolts were loosened. Threaded ½" Whitworth rod passing through holes in the chassis ends and nuts welded to the sub-frames allowed independent adjustment of each axle drive-chain tension from the countershaft. The seating arrangement was changed to allow the weight of the two 12 volt batteries to counterbalance that of the driver. This was best achieved by having a seat at each end and only one space, for the size 11 shoes, in the middle to minimise the overall length. The seats were constructed from ⅜" plywood on oak frames.

Pieces of this well seasoned limed-oak came from a huge wardrobe, which was given to me by a young couple who were our neighbours on the Ballycrune Road in the mid nineteen-seventies. She was an airline captain and he was a test pilot at Short Brothers & Harland. Many were the household articles, the tram brake-blocks and much else beside that were made from this source of oak timber.

Front and rear mahogany 'bows' were cut from pieces of salvaged sound-desk top. These were acquired when Alan and I were involved in the refurbishment of BBC Radio Foyle studios in Londonderry in the closing years of the Twentieth Century. The brake arrangements followed the pattern of Tram No. 4 including the source of the ratchet-and-pawl used. The fence at the garden gate was beginning to suffer from 'lack-of-tension' after this latest removal.

There were to be many occasions when the railway featured in the grandchildren's birthday parties, or at Easter, Halloween and Christmas festivities. On Christmas Eve 2006 a family party was held in the station. Train loads of grandchildren and parents were transported from the workshop to the station and mulled wine, lemonade and home-made mince pies were enjoyed before the show. A miniature theatre with remote control curtains had been constructed at one end of the building and the seating was five home-made wooden forms. With some hesitation due to a technical hitch the curtains opened to reveal three festive 'Coca-Cola' bears that performed the musical numbers 'Deck the Halls' and 'I'd Like to Buy the World a Coke'. These cute bears had been brought across the Atlantic Ocean by Andrew from Fort Lauderdale on a previous Christmas visit. See Plate 25.

Plate 25. The three bears

Watch the bears - visit this link:
https://youtu.be/MLpI6qwHLdE

There were several encores followed by 'speeches' before the audience were ferried back to the workshop terminus.

At a later date, a point was inserted just before the straight leading to the station. The branch led to a halt near the back door of the house where there were often barbecues of a summer evening. See Plate 26.

Plate 26. Tram No. 6 in the 'barbeque' siding

During this period the equipment in use was Tram No. 6, an eight-foot long bogie passenger car and a four-foot long planked bogie truck, all of which gave sterling service. The passenger truck was an attempt to maximise passenger-carrying numbers in comfort with minimum 'footprint' for storage in the workshop during the winter. This was very successful in that the two bogies could be 'dropped' by reaching through the slots in the sides and pulling 'R' pins enabling the lightweight plywood body to lift upwards. It was then stood on end with the two bogie units stowed inside the body. The planked bogie wagon was similarly constructed and could be either used for carrying 'goods' or pressed into service as a passenger vehicle for children with the addition of seats.

Over the years there had been a considerable number of wheels turned on my trusty ML7 lathe. In 5" gauge there had been four each for 'BBC', 'BAT', 'Friar Tuck' and a driving truck. In 7¼" gauge there were four each for Tram No. 3, Tram No. 4 and Tram No. 6. That is a total of twenty-eight. As will be seen in a later chapter there were to be many more wheels to be turned.

One memorable Halloween occasion was when my friend Stewart Clarke was driving a train load of visiting grandchildren towards the station in the dark. The children were enthusiastically waving 'sparklers' in the air and combined with the head and tail lights of the train it made a spectacular sight. The 'sparklers' were the invention of my wife, Ann. Concerned that the red hot variety could be dangerous, she came up with the idea of inexpensive 'Poundland' battery torches with strands of Christmas tinsel taped to the head. It would have been hard to believe in those halcyon days that three years later all this would be gone.

What happened was a combination of several different factors. The never-ending maintenance of our three-quarter acre hilltop property was becoming onerous with advancing years. With the onset of various aches and pains it became apparent that it could only get worse. I had never even considered moving to anywhere else. But with the passing of Ann's elderly father his detached bungalow was to be sold by the family. It was in a mature development and over the course of many years we had got to know the neighbours. The garden was very small but surely that was a good thing. We decided to 'downsize', sell our hilltop idyll and buy out the family.

Soon the 'Carricknadarriff Tramway' was gone. The purchaser of our hilltop residence had no interest in railways and so Tram No. 6, the eight-foot passenger coach, the four-foot open bogie-wagon and the entire track were loaded into a car trailer and disappeared down the lane. Like abandoned railways everywhere, all that was left were sad looking ballast paths where formerly the shining rails had run.

The Steam Dream

The year 2005 brought life changes. My elderly mother and my younger brother passed away within months of each other. The previous year I had just completed a fifteen-year mature student detour into academe resulting in the award of a doctorate. A great desire arose to build and operate a steam locomotive again. Long dormant memories of coal-smoke and steam issuing from the tall chimney of 'Sweet Pea', my previous adventure into the world of live steam, urged me on.

The decisions we take in life are influenced by many factors, some of them obscure. How I came to choose the steam locomotive, that would take me some ten years to build, illustrates this. My garden railway was 7¼" between the rails so the gauge was settled. I favoured narrow-gauge prototypes for their quaintness and ability to negotiate sharp curves that abounded on my line. Most miniature locomotives based of this type can be persuaded into a hatch-back car should the need arise to visit other tracks. But which loco? Well, when I discovered 'Marie Estelle' that was it. The name alone appealed to the romantic in me. The further I researched the background to this locomotive the more it bewitched me. Who was Marie Estelle?

It all began with Ollie Johnston. Ollie was an

animator, one of the 'Nine Old Men', who worked for Walt Disney and loved trains. Ollie's wife was called Marie Estelle. He acquired a full size narrow gauge Porter 0-4-0 steam locomotive in 1959 that was unsuited to the requirements of a Walt Disney theme park. The locomotive was dismantled and refurbished in the driveway of their home. The little locomotive coupled to a truck and a caboose ran in the backyard of their property. Ollie made a drawing, as animators do, of his engine incorporating some modifications to enhance its appearance. He sent the drawing to Don Young who was a miniature locomotive designer in England. Eventually a design for a 5" gauge version was produced called Marie E. after Ollie's wife. This proved to be such a success worldwide that a 7¼" gauge version followed. It was named Marie Estelle to distinguish it from its smaller sister.

This was the miniature locomotive that I had been looking for. It was relatively small and had a ride-on braked tender. Simple to build with bar frames and a wooden cab, it also had a big shiny steam dome, sandbox, and a bell. Up front was a tall chimney topped by an impressive diamond stack. See Plate 27. Studying the plans created visions of warm summer days, panting smoke and steam as Marie Estelle sashayed along narrow gauge track between verdant shrubs and trees. Sometimes it was winter with frosty nights, pulling 'Santa Specials' under the stars. Of course there were to be years of work, including the loco's four driving wheels and the tender wheels. These brought the total number of wheels turned so far to thirty-six. The idea was that the construction should take place in parallel with all the other short-term projects, especially when the weather prevented track laying and other outside activities. The copper boiler would professionally built but all the other parts of the locomotive were to be manufactured in the home workshop from castings and steel, brass, phosphor-bronze and copper materials.

Plate 27. Completed Marie Estelle

The first steaming took place one damp evening in 2014. To be honest it was a frightening experience. It was also rather unpleasant. With hindsight the home made chimney-top fan was inadequate for a 7¼" gauge locomotive and so the fire would not light properly, only producing clouds of foul-smelling coal smoke that wafted across one's face in the way smoke always seems to do. After a considerable length of time, the needle of the pressure gauge started to lift and 'boiling' noises were heard from the boiler. The battery-operated fan was removed and the 'steam blower' turned on. At last the gauge read 80psi, the regulator was opened and with an explosive 'snort' Marie Estelle took off like a startled horse. Initially, showers of condensate water, steam and black smoke blew out of the chimney straight into my long-suffering face. Visibility was reduced to nil due to steamed-up and water-splattered glasses. On the second lap of the circular track, things had settled down a bit until it was noticed that the boiler water level was getting low.

Now, all model engineers are aware of the 'bomb-like' potential of locomotive boilers that are allowed to run low on water. So an immediate stop was necessitated to concentrate on the job of hand-pumping water into the boiler. The fire was of course going strongly by this time and had to be controlled by opening the boiler fire door to reduce the draught. Then disaster struck! The water-feed hose to the boiler burst with a loud 'plop'. Emergency action was taken by dropping the fire, which almost immediately ignited the sleepers beneath. The combination of escaping steam and smoke from burning sleepers created a mini Dante's inferno scene that was a million miles away from earlier visions of 'sashaying through verdant foliage and Santa specials on starry nights'. It was all caught on video.

 The first steaming of Marie Estelle - visit: **https://youtu.be/ixgd-5qfQJs**

Since then the locomotive has been completed and an electric motor-driven water pump and a steam operated water injector have been installed. The coal fire-grate has been removed and replaced with a propane gas burner. These changes were carried to make operation of this little steam locomotive more suited to my small railway, where short runs are combined with long stationary periods. Only time will tell if 'the steam dream' can become a reality.

A New Beginning

The removal van was gone and silence reigned. The hall and passageway were lined on both sides with cardboard boxes, garden tools, chairs, unknown items under dust sheets and more and more boxes. The attached garage and all the rooms were similar, with only narrow walkways between the piles allowing access to a bed, the kitchen table or the oil-fired boiler. Even after many trips to charity shops and the local dump before moving, this was the result. Thirty-five years accumulation of 'stuff' in an old farmhouse and outbuildings was not easily reduced to fit into a modern bungalow. Never mind the sizeable workshop contents. Where was it all going to go?

To add to the difficulties my right knee had given up a week before we moved. Unfortunately 'keyhole' surgery provided no relief from the pain and there was a nine months waiting list for TKR or total knee replacement. Although care had been taken to label boxes with helpful notes on their contents, the notes were never detailed enough to find, for example, a certain length and diameter of wood screw. Much frustration ensued when trying to do the simplest job like mounting a towel rail. Where was the hammer, the 7mm masonry bit or the plastic wall plugs? The frequent scrabbling in the boxes gradually reduced their contents to a 'dolly mixture'. It was difficult to see what was in the bottom of the box anyway, since none of the torches could be found. No uncluttered horizontal space existed to up-end the container and sort the contents. Still, I suppose we were lucky. On asking the removal boss, "Is this the worst case you have ever seen?" He replied, "Not at all, on the last job we had to leave an uncovered pile of furniture on the front lawn in the rain."

A drastic situation can only be remedied by drastic measures. Our small touring caravan was sitting on the poorly drained, and hence soggy, front lawn. The short driveway only accommodated the car. A plan emerged to create a 'runway' of flags from the road on the other side of the bungalow for the caravan. That would fix the parking, what about the inside of the bungalow? Well, the garage was accessed by a fire-door from the hall through a fire-wall consisting of concrete blocks laid on their flat. It had light ceiling joists partially covered with chipboard sheeting, suitable only for storage. This had been reached by a substantial shallow stairway suited to my wife's elderly parents. Three improvements were needed: (1) Remove the stairway and replace it by a smaller steeper one. (2) Support the lightweight joists by two massive wooden beams. (3) Fit a new fire door to give access to the area above the living quarters of the bungalow. A friendly small builder known as 'Stevie' gave an acceptable quote for the job. Several days later all was complete and a whole new world was opened up!

The first opening of the upper fire door revealed a vast area with good headroom. But what ensued there will be described later. The space above the garage was going to be the new workshop. After many years of working in arctic conditions in stone outbuildings at our previous home, some consideration was to be given to comfort. Once the new stairs were installed the floor was completed with further chipboard sheets and a ¾" ply trapdoor to cover the stairwell. The door was hinged to a joist attached to the wall and counter-balanced by an old sash window weight with its sash cord passing over a pulley hung from a purlin. Two 'Velux' windows were installed in opposite sides of the sloping roof, admitting daylight and providing ventilation. They each gave a 'window on the world' making one aware of the changing light and weather. The two sloping parts of the roof and the horizontal joists were sheeted with plasterboard and a generous thickness of glass fibre insulation put in place. After plastering, walls and ceiling were painted with white silk emulsion. A ring-main circuit for 13 amp sockets and fluorescent lighting were installed. Finally, vinyl sheet was laid that had been a temporary kitchen floor covering.

The new workshop was now ready to receive its equipment. Getting it up there proved surprisingly easy after purchasing a mains powered electric hoist that was suspended from the purlin over the trapdoor. It had a half-ton lifting capacity, and so even the ML7 lathe on its wheeled stand was rapidly raised and put in place. With my wonky knee confining me to 'light duties', I was most fortunate to have a very generous friend, Ken, who installed the windows and sheeted and plastered the ceiling. With the workshop equipment gone to a higher place, the garage was now able to accept much of the household items that had been lining the rooms and corridors. Another huge benefit was that all the tools, fixings and DIY equipment now had workshop storage places and were easily accessible.

Within two years of moving house, thoughts began to turn again to garden railways. Casually leafing through an old copy of 'Model Engineer' a photograph caught my eye. It depicted a small 7¼" gauge tram-like locomotive, with sit-in driver, pulling several passengers. It was called the 'Pocket Rocket'. Now, this vehicle interested me very much. Although I was building a Marie Estelle I wanted something that could be completed quickly and at a minimum of expense. If there were ever to be a new 7¼" gauge railway at our downsized home it would have to be very small and employ sharp radius curves. New equipment would have to cope with this. The 'Pocket Rocket' looked to be around three feet long, which was about the minimum size that would comfortably accommodate my six-foot frame. It seemed a good design concept that the driver's weight would provide the necessary adhesion for passenger hauling, so that the vehicle itself could be relatively light for handling.

To minimise weight, a mostly wooden construction was envisaged, similar to a nineteenth century narrow-gauge electric tram. A modular design was adopted that would allow disassembly for stowing in the back of the hatch-back car for transport to Cultra. Was this project going to be expensive? Happily, the answer was no! The reason for this requires a diversion into a sort of engineering fairy story.

Once upon a time, at a land-fill site far away, a model engineer saw a trailer load of electric bicycles about to be dumped. Upon enquiry, he was told that they were brand-new but since they lacked a CE mark they could not be legally sold for use. The bicycles were diverted to the model engineer's garage and everyone lived happily ever after.

I acquired two of these 'machines' and proceeded to dismantle them for parts. Each electric cycle comprised the following items that were retained: A ⅓ hp 24 volt traction motor; drive chain and sprockets; two 12 volt batteries; electronic control box; twist-grip throttle control; key-switch with key; fuse-holder and fuse; head, tail and indicator lights; electric horn with button switch and the wiring harness for it all. Ironically, during the deconstruction process a hidden CE mark was observed.

My first thought was to build a compact 3½" gauge locomotive that would transport easily to the superb raised track at Cultra. Several trips with a tape measure to the Folk and Transport Museum, enabled me to measure up the full size three-foot gauge 'Phoenix', long retired from the County Donegal Railways after nearly a quarter of a million miles in service. Construction went ahead and the finished locomotive proved to be a great success, easily pulling four passengers effortlessly around the circuitous track. See Plate 27. The first trial run of the bodiless chassis was 'electrifying' to say the least. It took place on one of those summer afternoons that one dreams about in the depths of winter. There were few fellow model engineers about as I climbed aboard the passenger truck behind my 'brassy' (still unpainted) Phoenix. Key-switch on, throttle gingerly opened and we were off and swishing along at a spirited pace. I never did get to open the throttle fully as this high-speed run was probably already the fastest, most exhilarating trip on rails that I have ever experienced.

 Experience the run - visit:
https://youtu.be/Vh1dCcx9XK0

Plate 27. Completed 3½" gauge Phoenix at Cultra

This success led me to believe that the second set of parts could be used for my new 7¼" gauge tram. This was my fourth tram build and I intended to incorporate further lessons I had learned from the other three. A 15" wheel-base was adopted as a compromise between the ability to negotiate tight radius curves and loading stability. Chassis construction was to be very different from previous efforts. Instead of conventional steel slotted side plates with sliding axle boxes, three-foot long timber pieces 2" x 1 ¼", sawn from standard 4¼" x 1¼" PAR door framing were employed. Eight ⅝" ID pillow-block pressed steel bearing housings were purchased. These were the least expensive option for medium duty use that suited incorporating spring suspension. Four pairs of holes at 68mm centres were drilled in the side frame timbers to suit the bearing mounting centres. There was some ½" OD thick-wall brass tubing in the workshop stock, so 2" long pieces were used as sleeves in the wood for the sliding bolts. The holes in the wood were first drilled 13mm and then reamed out using a piece of the tubing with the surface roughed with a coarse file. The sleeves were then pressed into place using the drill press. Suitable bolts were found to be 8mm x 90mm. These were dropped through the counter bores in the ¾" ply decking plates, the brass sleeves, the turned centring cups and the springs. A reduced diameter 8mm nut was then screwed up until just inside the spring, the feet of the bearing housing shells were then retained by 8mm locknuts. See Plate 28. The springs, ¾" x 12SWG, were cut to length from 6" lengths of compression springs bought at the local hardware shop. The corner of the wheel of the workshop grinder was used for this and then the cut ends were flattened on the face of the wheel.

Plate 28. Tram No. 8 chassis and 'works'

Two wheel-and-axle sets, previously made for a 5" narrow gauge driving truck, were machined to 7¼" narrow gauge profile. The wheels finished to 2½" tread diameter, which was rather small by normal standards but it was felt that along with the short wheelbase the tram should cope with my tight radius curves. An essential requirement was to have the tension of each drive chains independently adjustable. To achieve this, a folded 19SWG galvanised steel motor-tray was made that was fitted with skateboard bearings mounted in turned housings at one end for the countershaft and slotted holes for the motor at the other. In turn the motor-tray was carried on slotted mounts and could be tilted by 2BA jacking bolts.

The modular construction of the tram consisted of this motorised wooden chassis, a bolted on body, rear narrow-gauge style buffer/coupling and a lift-out battery tray. The tram body was designed around a tunnel between the driver's feet to accommodate the 'works', footboards to each side, a seat behind and curved metal panels at both ends.

Apart from the 18SWG aluminium end panels, the rest of the construction was ⅜" and ¾" plywood on 2" x 1" PAR laths. Four 3" x 6mm coach bolts secured with wing nuts fasten the body to the chassis via rubber door-stop spacers, thus allowing quick removal. This feature later proved useful in that ground clearance could easily be adjusted by means of plywood 'washers'. A removable trap door on the top of the centre tunnel gives access to install the 24 volt battery tray. Oak brake shoes operating on two wheels are applied by a tram-type handle via a chain that winds round the vertical ratchet-and-pawl controlled shaft. The cast iron ratchet wheel was part of a fence tensioning fitting from the hardware shop and the pawl is a length of 12mm x 3mm, foot operated, to release the brake. Towing a passenger car was catered for by a sprung combined buffer/coupling. It is removable by pulling a pin, allowing the tram body to lift off. Polished brass house number 8s were screwed to the end panels and a brass headlamp provided for any night trundling in the garden shrubbery. All the electrical control equipment was housed in a wooden box with a brushed aluminium facia. The only extra to the parts that came with the electric bike was a reversing switch; cycles apparently do not need to go backwards!

Since there was still no track laid at home, it

was decided to try out the portability and performance of the completed tram at the MESNI track before painting. Unfortunately, when fitting the battery tray the flying-lead spade terminals were connected in reverse. A loud 'phut' from the control box announced that the 20 amp fuse had blown. There was no spare to hand. Luckily a small piece of litter saved the day. A discarded foil wrapper rolled round the blown fuse allowed the current to flow again and tests commenced. These went so well that several model-engineer observers were invited to entrain. Half the 7¼" gauge circuit was completed when all came to an embarrassing halt. Fuse again! Still, it was a lot to ask of a cycle-motor with a chocolate wrapper fuse. The tests were adjudged to be a success. In the succeeding months and years many other cycle-motor powered locomotives were to appear on the raised 5" gauge track at Cultra.

Once home, the wooden body received several coats of varnish. The end panels were sprayed blue as a result of finding a forgotten can of that colour on the garage shelf. One of the early influences on my interest in narrow-gauge electric trams was the 'The Giant's Causeway Tramway'. This three-foot gauge hydro-electric powered line ran from Portrush to the Giant's Causeway via Bushmills. The cars had 'toast rack' seats with gracefully curved ends, a feature that I copied. And thus was born 'Bushmills Bluey', carrying the No. 8. See Plate 29. Whilst the tram was being built in the workshop during the winter of 2009 occasional forays were made into the small back garden with a surveyor's tape measure. At first the possibilities looked depressing.

Plate 29. 'Bushmills Bluey'

Rocklands Railway

With the success of 'Bushmills Bluey' came a renewed desire for another garden railway. Even though the 7¼" gauge 'Windy Hill Railway' had been sold off, quite a few keruing sleepers and lengths of aluminium rail remained in stock. These had been kept 'just in case'. A preliminary survey of the small garden was dismaying. An 8' x 6' garden shed standing on concrete pillars had been positioned such that it decreased the available space even further. Another snag was a fall of seven inches towards the right hand back corner where the shed was standing. The first consideration was just how large a radius of track could be accommodated. Using a surveyors tape measurements were taken and an A3 drawing produced showing the outline and position of any immovable objects. Whilst carrying out the survey it was noticed just how soggy the lawn was, thus indicating poor drainage and a further problem. A fifteen-foot radius circle of track was found to be the maximum that could be fitted into the available space. But what to do about the fall in ground level? Various schemes were doodled on the backs of envelopes. Options included all-welded steel bar track similar to that used for the 'Windy Hill' 7¼" gauge railway that needed only widely separated supports or perhaps aluminium and wood-sleeper track laid on some kind of wooden decking. Ultimately, the solution came from my wife who was getting tired of squelching across the 'lawn' to the washing line.

"Why don't we make a circular patio big enough to accommodate your railway? It can be partially flagged for pot-plants and a sitting area and the rest stoned to give good drainage."

There was now a level platform on which to lay conventional wooden-sleepered aluminium track. Fortunately my home-made rail bending machine still existed. See Plate 30.

Plate 30. Home-made rail-bending machine

The keruing sleepers were jig drilled to suit ⅛" over-gauge curved track. A total of 984 pan head 5mm x 25mm long stainless steel screws into 164 sleepers were required for the thirty-foot diameter circle of track.

 Experience the new track - visit:
http://youtu.be/dZnGEQiv3Ss

Although a small child could sit on the carpeted tunnel in front of the tram driver, it was obvious that a passenger truck was urgently required. Experience of building 'Bushmills Bluey's' chassis suggested that bogies for a truck could be constructed along similar lines. The 'Ecobogie', see Plate 31, measures 18" long by 11⅞" wide. It has a ¾" plywood deck screwed to 1¼" wide x 2" high side pieces, with the bearings attached in the same way as the tram chassis. The centre of the deck has an upstanding ½" diameter swivel pin with an 'R' pin retainer. There are two 1¼" square x ¼" thick nylon rubbing pads screwed to the top of the deck to each side of the swivel pin. For towing, each end of the deck has a 12mm x 3mm drawbar fitted with a 6mm hexagon headed coach screw into a drop-block 3¾" x 2" x 1" deep. The wheelbase is 10". Wheels are 3¼" tread diameter x 1⅛" thick fitted to 1⅛" diameter axles with 'Loctite'.

Plate 31. 'Ecocbogie'

The truck body is a simple tray with steel plates screwed underneath that rest on the rubbing pads of the bogies, the swivel pins passing through central ½" holes in the plates and wooden body. Foot rests are hinged to the sides such that they can be folded upwards to reduce storage space. The last item in the assembly is a longitudinal seat for sit-astride passengers that slots into the tray. See Plate 32.

Plate 32. 'Ecocoach'

Fortunately the Rocklands Railway track is generally level with no steep gradients. Although there is a long straight along the side wall of the bungalow, there are several sharp curves and a continuous circle in the back garden. With a heavily loaded train, comprising a tram and two coaches, there is considerable cumulative flange friction from all those wheels, especially when the rails are dry. A trial with a can of silicon spray applied to the points and the curves produced a much quieter and easier ride. 'Bushmills Bluey' has now been fitted with small-bore nylon tubing that directs the silicon to the running face of the rails. A quick trip round the track before setting off with passengers is all that is required to enhance the experience and reduce track wear. It seems that track lubrication is common in full-size railways too.

Points and Signals

A circle of track in the back garden was a considerable improvement on having no garden railway. However, an outing was now a matter of going round and round like 'the wheels on the bus'. The thirty-foot diameter track was laid on flint-stone ballast that was 2" lower than the top of the concrete kerbing forming the retaining edging. Since the wooden sleepers were 1" thick the bottom of the rail was only 1" below the edging top. So the daring idea was born, raise the ballast locally by 1" and points could be constructed so that the base of the rail sat on top of the kerbing. No damage need be caused in escaping from the circle.

Making points is not easy for the novice. Even after constructing seven sets over many years, admittedly for various rail sections, it is still a challenge. Shaping the taper on the point blades is best tackled by milling using a simple jig on the lathe cross-slide or milling machine table. Excellent instructions for point construction can be found in the '7¼" Gauge News'. It is well worth joining the '7¼" Gauge Society' to use their extensive archive. However, it is still possible to construct points the hard way using ingenuity, a hacksaw, an angle grinder and even an electric plane where the rails are aluminium.

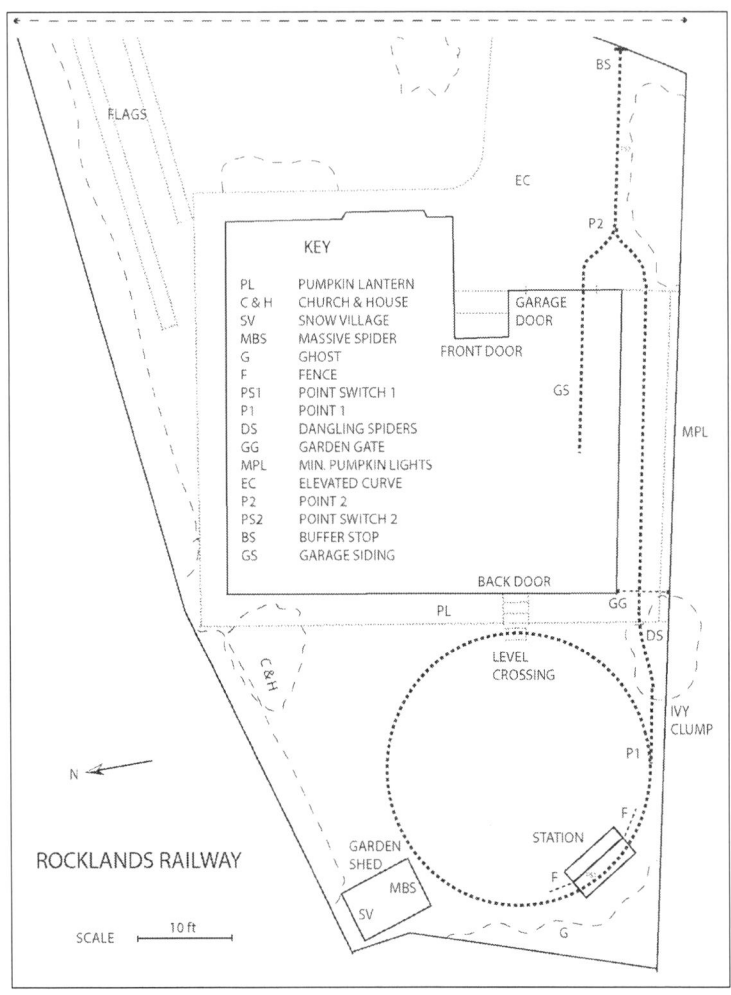

Plate 33. Rocklands Railway

Some surgery was needed to the existing track to insert the point as the joins had been staggered to avoid '3D bitting'. See Plate 33 for the references in square brackets. Immediately after the point [P1] a chicane was required to achieve minimum track clearance from a gnarled ivy clump. The overhanging fronds of which were to later provide a spooky place for dangling spiders for the ghost train at Halloween. Proceeding through the gate [GG] along a straight run of track, laid on existing 3' x 2' concrete flags at the side of the bungalow, brings us to an elevated left hand curve [EC] leading to the tarmac in front of the garage. It was intended that this curve and the rest of the track and point [P2] could be easily and quickly assembled and disassembled. On special occasions it could be brought out from storage in the garage and set out in a matter of minutes. The curve needed to be elevated to cope with a difference in levels between the flags and the tarmac drive. Not wishing to have to build up the track with numerous different wooden wedges, which could get lost, it was decided to jack the sleepers with 8mm coach bolts. These bolts were inserted in threaded holes near the ends of the sleepers, large dome head in contact with the ground, and locked in place with a nut. The dense keruing hardwood of the sleepers threaded well with an 8mm tap held in the chuck of a battery drill. Adjustment of height is facilitated by use of a spanner on the square shoulder of the bolt. Joining track panels together in alignment was achieved using 1" x ⅛" flat alloy strip secured to the sleepers by two stainless steel wood screws at one end and a 6mm coach screw with a wing-nut at the other.

Not wishing to have to cut clearance slots in the bottom of the up-and-over garage door (likely Domestic Authority disapproval), a short section of track under the door was made removable. Thus the complete train could be parked on the garage siding

[GS] in the event of rain or during the night with the garage door closed.

Plate 34. Electric point mechanism [P1] with manual lever override

Not long after the points went into service it was realised that a great improvement would result from being able to change direction from the driving seat of the locomotive. Remote mechanical operation was ruled out, especially for the removable point [P2], as being too complicated. It was decided that electrical remote control would be better. Car door remote-locking mechanisms seemed to offer a good solution. They operate from a 12 volt battery, are small and powerful, have built in limits to their movement, can be mechanically overridden and, lastly, are very inexpensive if ordered direct from China on eBay. Point 1 already had been fitted with a weighted lever and this was retained for manual operation, a feature that seems to fascinate younger railway staff. See Plate 34.

Since it is a trailing point when travelling clockwise round the loop, the operating switch [PS1] is in a fixed position with its 12 volt battery (lead-acid burglar alarm battery) in the station. It can be easily reached rom the locomotive with or without a train of coaches when travelling in either direction. The 10 amp DPDT toggle switch is wired as a reversing switch, centre off and momentary make in either direction. See Appendix I. My switches were obtained from Maplin Electronics and can be fitted with a rubber boot when used in an exposed position. Point 2 switching via [PS2] required a different approach. The switch needs to be moveable so that it is within easy reach from the locomotive when it is at either end of the train. Switch [PS2] was fitted to the lid of a plastic box with the 12 volt battery inside. To make the switch easy to reach it is raised 10" above a heavy wooden base by a length of brush shaft. Twin flex, long enough to allow for the change of position, is connected to the point via car-type bullet connectors. In use, the box can be turned around so that the switch toggle movement indicates the correct road. This prevents surprises when night driving!

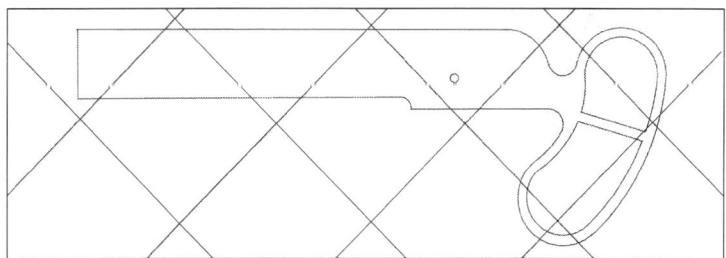

Plate 35. Signal arm outline

I always wanted a signal. There is something attractive about the way the arm clanks into position operated by some unseen hand. A favourite haunt of mine is the County Donegal Railway Museum in

Donegal Station and they have a very nice signal in the station yard. A photograph was taken square on and the shape traced out and diagonal guidelines superimposed in Adobe CS2 Illustrator. See Plate 35. The image was then printed in parts on three sheets of A4 paper. When the paper sheets were trimmed the diagonal lines assisted in lining up the joins. The paper was stuck to ⅜" ply using Prit Stick and the signal arm complete with lens apertures cut to shape with a scroll saw. Next the aperture area was placed on 18SWG brass sheet and a felt-tip marker run around it and the openings to mark out a ⅜" wide frame. This too was cut out with the scroll saw. After painting the arm, red and blue 'glasses' were sandwiched between the brass surround and the plywood and retained by No. 2 x ⅜" brass countersunk screws. A 1½" square x 6' high wooden post, painted white, held the signal arm and operating box aloft. It was surmounted by a finial, an upside down plastic ground post from an LED garden light, also painted white. See Plate 36. The base, underneath, was made from 'slater's lath' screwed to a 400mm garden flag. All that was the easy bit.

On the Rocklands Railway, in common with many other garden railways, the driver is the signalman, the guard and the ticket collector so signal operation would have to be remote and portable. The first attempt worked but was too slow. A small 12 volt motor with reduction gearbox was mounted in a 6" x 4½" x 3" waterproof plastic box with the signal arm fitted to the protruding ¼" diameter motor shaft. A miniature arm inside the box was arranged to operate two micro switches wired as limit switches. See

Appendix I. Control of the motor was achieved by use of a 27MHz radio control transmitter and receiver purloined from a now-disused model yacht. The receiver drove the motor via a 'Mtronics Viper Marine 15' motor controller. A 9.6 volt battery, consisting of eight AA rechargeable cells in a nylon holder, with 3 amp fuse went inside the box. An on/off slide switch operable from outside and two battery charging posts completed the installation. The signal arm could be raised or lowered from anywhere in the garden or from a locomotive but it was too slow and there was no 'clank'!

Plate 36. Scaled copy of CDR signal

After a summer's use when it was found that an operating signal greatly added to the interest of running such a small railway, it was decided to rebuild it. Thoughts turned to the use of a car door-lock actuator when it was realised that the action was similar to that of the existing points. There would be two improvements; the action would be much quicker and the bulky 27MHz transmitter would be replaced by a small key-fob. Two key-fob transmitters, the receiver and four actuators with wiring harness were obtained on eBay for £18 post free. There was a problem at first due to the force needed to move the unbalanced arm. When a counterbalance weight was fitted to alleviate this, a new difficulty arose in that even a small gust of wind was able to move the arm out of position. Looking at the full-size signal post provided the idea for a cure, use of an over-the-centre weighted arm as in the prototype. And there was a 'mini clank'.

As the long dark evenings approached there was a need for a signal lamp. Apart from the operational requirement of being able to see the set of the signal in the dark, it would be attractive looking out of a window of a December evening. Electric lighting was ruled out. I wanted to perform the duty of lamplighter as well as all the other roles. So it was decided to model an oil lamp but light it with a 'tea light' for simplicity. Since the lamp would have to be out of doors in all weathers, construction in non-ferrous metals was indicated. Happily, there was plenty of copper sheet in the workshop store as the result of replacing a direct hot water cylinder in the dim and distant past. Another useful find was that of a bullseye lens, one half of a lantern slide-projector condenser lens.

Plate 37. Signal lamp

The domed aluminium base of a Coca-Cola can provided the reflector behind the flame. To make it easy to light and replace the 'tea light', its holder was mounted on a swinging door. The parts were soft soldered together and all screw and nuts were of brass. See Plate 37.

There may be thoughts of linking the signal to a 'Raspberry Pi' computer that will coordinate the chuffing and whistling of trains to the signal operation. Or was that just 'hedge-conditioned' home brew thinking?

Celebration Coach

As Christmas 2013 approached, it was becoming apparent that there was a need for an extra coach that could accommodate the less flexible railway travellers. It was decided to make it as spacious and comfortable as possible. Therefore a five-foot long bogie-coach with a foot well and a roof to provide some protection from the elements was suggested.

Towards the end of November a trip was made to the local steel stockist and a length of 40mm x 40mm x 3mm angle plus one of 40mm x 3mm flat black iron were purchased. Fortunately my 'Mini Mig 130' was happy to weld this weight of material. Using a mitre band saw the bits were soon cut to length and the under frame welded up. The construction of the two bogies followed the economical design developed for the sit-astride coach as described earlier. This time though, there was a quandary regarding the wheels, all eight of them! For some time there had been four slices of 150mm diameter steel bar languishing in the store cupboard from an abandoned project. There was also a slice of 100mm diameter stock. What to do?

Following the minimum cost and remembering that time is not money when one is a pensioner; a labour intensive solution was adopted. The 100mm round was placed on each of the four 150mm ones and a circle drawn round it with a permanent marker. A suitable piece of timber was gripped in the saw vice and the large round clamped to it in a horizontal position so that the blade would saw off a segment close to the line. A series of cuts removed most of the excess material, leaving little to be machined in the lathe. The others were treated similarly. Although this seems like a tedious process, there was a sense of satisfaction since other work could proceed as the gravity-fed saw got on with the job and switched off automatically at the end of the cut. Admittedly there were some occasions when the saw blade got pulled off its rollers repeatedly and all was not 'sweetness and light'. There were now five wheel blanks but three more were needed. A trip was undertaken to a reclamation yard hidden deep in rural County Down. Beside one of the many mounds of vintage materials was a pile of rusty steel bar off-cuts on a pallet. For a consideration of £5 a length of 100mm diameter steel was acquired, enough to slice into three more wheel blanks using the trusty band-saw.

Each blank was gripped in turn in the four-dog chuck of the lathe and roughly centred. The back of the embryo wheel was faced and then drilled and bored ¾" diameter. The bore was made a neat fit for a previously turned mandrel, made from a broken 2MT drill. The flange outside diameter was then turned to size and the edge radiused. All wheels were completed to this stage. Each wheel back was now drilled and tapped M6. A catch-plate was then mounted on the headstock spindle nose and the taper mandrel inserted in the bore. A wheel was slid onto the ¾" stub protruding from the nose and secured by means of a M6 bolt and washer through the catch-plate. The first operation was to face the front to give a finished 1⅛" wheel thickness. Then with the lathe in back gear and using the fastest of the three speeds, one angled tool setting allowed the 2 degree tapered tread and the 20 degree flange wall to be machined. If tool chatter was experienced at the flange root (as it was), the motor was stopped and a finishing cut taken by pulling round via the belt. All that remained was to radius the flange tip with a file and chamfer the outer edge of the tread at 45 degrees.

Now, eight wheels with inner and outer surfaces to be faced require an awful lot of tedious handle twiddling. With a touch of arthritis in the joints it can be a painful process, in fact a job that was 'hard to face'. Thus some thought was given to a power cross-slide.

At first a battery-electric drill seemed a very suitable power source. It had a low speed gearbox with electronic speed control and reverse switch all built in. However, early experiments soon showed up a number of difficulties. Firstly, the battery drill was heavy and required a substantial and clumsy mount attaching it to the cross-slide of the lathe. Secondly, the speed control and on/off switching was via the trigger and it was found difficult to accurately depress it to obtain consistently the very low speed required. Indeed each time a facing cut was tried, the tool 'dug in'. Thirdly, reverse control was achieved via another small lever associated with the trigger. The final problem was how to disengage the drive to allow 'hand tweaking' of the tool when required with such a clumsy set up? A much more elegant solution was required. See Plate 38.

Plate 38. Motorised power cross-slide unit

The problem was solved via a small 24 volt DC motor with a low speed gearbox from the 'might be useful in the future' shelf. The 2BA cap-head screws attaching the cross-slide bearing end plate were removed and replaced by ¼" diameter rods with the ends threaded 2BA. Two holes were drilled in a piece of oak timber (from the old wardrobe) at the appropriate spacing to allow it to slide freely on the rods as a motor carriage. The outer ends of the rods were held in another piece of oak acting as a spacer bar and mount for a forward/centre-off/reverse tumbler switch. This was fixed in place by two ¼" Whitworth bolts into holes tapped in the oak. A simple two part dog-clutch, one part mounted on the gearbox spindle and the other replacing the cross-slide ball-handle, gave an easy engage/disengage action by moving the motor carriage. Thus freeing the cross-slide for manual action when required. These two parts were: (1) A sawn off female hexagon piece from an old driver handle drilled to fit the gearbox spindle and tapped 4mm for a grub screw. (2) A hexagon driver bit threaded ¼" BSF with locknut attached to a 2" length of ⅜" x ⅜" steel bar that replaced the ball-handle of the cross-slide. The motor and tumbler switch were wired via a small micro-switch as seen in Appendix I. This was mechanically adjustable to act as a limit switch. At first it was intended to employ two limit switches but it became obvious that really only one was required. This was due to the fact that on the outward travel from the centre an overrun simply caused the cross-slide feed screw to leave the feed nut, and the cross-slide to stop. A twin-flex cable to a retired drill battery on a shelf completed the assembly. The current consumption being so low, around 30mA, the battery lasted for days of facing.

This simple piece of equipment was made up very quickly, and after confidence was acquired in its reliability, enabled other jobs to be undertaken while machining was in progress. Occasionally, it was a pleasure to have a cup of tea and watch the swarf accumulate! An added bonus was the superior finish achieved by the steady action of the motor compared with manual operation.

With the bogies completed progress could then be made to a rolling chassis. The load bearing floor of the coach was made from ¾" exterior ply sitting inside and screwed to the angle chassis with countersunk-head wood screws. This was possible as the weight of the passengers on their seats was sitting on top of the plywood and not on the chassis, thus making a neater job.

A 3mm steel bearing plate with a ⅜" hole for the bogie pin was screwed to the bottom of the ply, and a small circular steel plate, also with a ⅜" hole, was fixed above. I could never have imagined the scenario that occurred just before Christmas 2013 as a result of fitting that 'small circular steel plate, also with a ⅜" hole'. In the dark of night, a British Airways long-haul pilot was to be found lying full length on the ballast with a torch, peering at the rear bogie of the 'Celebration Coach'. It persisted in derailing as it passed over the point and he was trying to help me find the cause. That turned out to be due to a lack of 'compliance'. In other words the pin was too good a fit in the ⅜" diameter hole, thus not allowing a little lateral freedom for the bogie to follow irregularities in the track. The problem was solved by slightly increasing the size of the hole.

As far as possible, locomotives and rolling stock for my railways have been constructed in such a way that they can be dismantled quickly and easily. This is for two reasons: (1) So that the individual parts are of a weight than can be lifted for storage or maintenance, since it is amazing how everything seems to get heavier and further away as one gets older. (2) To make bulky items easier to store during periods of disuse, such as during the winter. Following this principle, the bogies can be detached from the chassis by pulling 'R' pins. Each of the two seats which include an end panel and two side panels, all in ⅜" birch ply on 2" x 1" frames, can be detached by removing four 6mm wing nuts. Lastly, the roof can be lifted off its supporting poles by slackening four 6mm wing nuts. The four poles then simply lift out of their corner supports. The roof consists of a ¾" thick softwood frame and curved ribs covered by hardboard textured side up to emulate canvas, pinned and glued to it and painted black with floor paint. All the other wooden parts are varnished.

Plate 39. 'Celebration coach' under the ivy

It was intended that this conveyance, now named the 'Celebration Coach', would feature in various railway activities during the year. With this in mind, four 13" high oval frames were cut out from 3mm ply and screwed to the seat side-panels with roundhead brass screws. A source of colourful artwork was found in large gift bags, sold in stationers and gift shops. The front and back of a bag provides two panels and the surface seems to be reasonably moisture proof. Christmas, Easter, Halloween and birthday events can be featured by changing the artwork behind the ⅝" wide frame rims. See Plate 39. A small 'tea light' lantern fitted with a red gel as a tail light can be clipped to either end panel. To add to the 'Pullman' luxury look, it is intended in the future to fit a swing away table and a shaded table lamp.

Celebration Trains

With the approach of the Autumn Equinox and the shorter days there is sadness for the loss of the long summer evenings. However, every cloud has a silver lining, and despite slippery leaves on the track it is a season of new possibilities. A narrow-gauge garden railway really comes into its own when it is dark. Especially so, if the clearances are tight and the dim headlight allows only glimpses of objects and foliage as you are guided along serpentine paths. Surely just the circumstances for a 'ghost' train?

Remembering amusement-arcade ghost train rides from boyhood suggested various ideas. Sounds, physical contact and very limited vision seemed to be the effective requirements. But how to achieve an exciting automated ride that would allow the train driver to concentrate on real hazards in the dark of night? Gradually a plan unfolded. As the train travelled the route it should trigger 'scary' events, yet since my track included a small circle it would be more effective if the 'events' could be random. After some thought it was decided to make one of the new Raspberry Pi computers the heart of the gubbins. It was chosen for two reasons. It has input/output connections that can be programmed to work with external apparatus and it is inexpensive. To switch events, the computer requires only a momentary 'on' pulse when the train is at the right place on the track. Burglar-alarm door/window sensors comprising a magnet operating a reed-switch proved suitable for this purpose. A reed-switch block screwed to a sleeper was wired, using alarm cable, to the computer in the garden shed. To operate it a magnet block was attached to a swinging arm under the tram, arranged to just clear the switch-block on the sleeper. See Plate 40.

Plate 40. Magnet block swinging arm

The swinging arm was to cater for the possibility of small obstructions on the track. Positioning of the track switch was determined by deciding when the passenger truck would be opposite .the 'happening'. Three 'scary' events were arranged, all with sound: (1) Small face-brushing dangling spiders under overhanging ivy briefly illuminated from above. (2) An illuminated ghost. (3) A large green backlit spider in a doorway. The computer programme was written by my ever-helpful son, Alan, for another of Dad's wild schemes. The Raspberry Pi listing is reproduced in Appendix II.

It was arranged that there were more sound effects than events so that they appeared to be random. The ghost was simply a white pillowcase with a balloon head inside supported by a brush shaft and illuminated from below. To allow the Pi to switch the various lights directly, the current had to be kept low. Inexpensive multi-LED clusters were purchased from the local 'Pound Shop' and wired from the computer using two cores of the same cable conveying the switch impulse. Where a higher wattage light was required, as in the backlit spider, a small 'Maplin' relay was used.

Audio distribution was achieved by means of an inexpensive FM transmitter sold for transmitting music from an iPod to a car radio. The transmitter was plugged into the 3.5mm Pi audio output socket in the garden shed. Three old FM radios were tuned to the appropriate frequency and placed in sealed plastic bags at the respective locations. To avoid running out of battery power, mains extension cables were run to the radios. These were plugged into a RCD socket in the garden shed for safety.

Five sets of illuminated 'pumpkin lanterns' were strung along the fence at the side of the bungalow. This feature and the location of the other 'events' are marked on Plate 33. A much admired additional feature was Alan's carved and candle-lit pumpkin lantern guttering in the night breeze. See

Plate 41.

Plate 41. Alan's carved pumpkin lantern

Let us take a trip on the ghost train as a visiting child on Halloween night. Entering the garage by the side-door from the house, 'Bushmills Bluey' sits coupled to its train of coaches. The panels of the 'Celebration Coach' display witches on broomsticks flying over orange moons. The subdued light is from a set of miniature orange pumpkin lanterns, Bluey's dim headlamp and a red tea-light powered tail light. After climbing aboard with several other children and Sophia's (youngest granddaughter) intrepid mother, several loud 'dings' from the tram's gong announces departure. We reverse out into the night to an accompaniment of clicks and clacks from the many wheels passing over the rail joints in the track and points. The train halts with the last coach perilously close to the roadway. The driver operates the track-side switch for the electrically-controlled point and with a resounding 'ding' we set off along the side of the house with strings of tiny lit-up pumpkins on the fence to our left. Suddenly, under a tunnel of overhanging ivy, hairy spiders dangle above us glimpsed in an unearthly light and accompanied by the sound of clanking chains. After a short distance curving to the right, the train passes through a dimly seen station. Out of the darkness appears a fluttering ghost with a drawn-out groan.

A little further on, a piercing scream emanates from a doorway in which a huge spider is silhouetted by a ray of green light. Soon, we pass a silent scary pumpkin head, eyes and mouth lit with a guttering candle. Now, after rumbling over the level crossing at the back door of the house, illuminated by a stream of light from the kitchen, we pass over points and enter the station. We stop. The driver operates the points switch. A warning 'ding' and the train reverses back along the side of the bungalow, stopping at the road, another 'ding' after the points are set for the garage and shortly with clicks and clacks from the wheels we are back under shelter. 'Apple ducking' commences...

Soon it is time to remove the Halloween art from the coach panels and think of pictures of Santa and his reindeer. After the success of the ghost train the grandchildren, and some of the more imaginative adults, would be expecting something special. As previously mentioned, a door from the upstairs workshop gives access to a huge area above the domestic parts of the bungalow. When I first opened the door it shouted 'theatre' to me! Although this was the year of the first knee replacement, by adopting various techniques, such as sitting on a garden kneeler or lying full length on a temporary sheet of flooring, it was possible to floor the loft with 8' x 4' x 19mm chip board. Because it was anticipated that the children would want to jump about a bit and the main area was quite unsuitable due to lightweight joists, a stage was constructed. By suspending heavier joists from an existing RSJ at one end and wall-mounted joist hangers at the other, an area independent of the downstairs ceiling was created. The end wall opposite the entrance door was large enough to accommodate a king-sized cotton bed sheet as a screen. The top of the 'screen' was pinned to a roof tie and the bottom to a length of 2" x 1" lath. Since it was 'for the grandchildren at Christmas' my long-suffering wife agreed to the purchase of an HD video projector.

So what was missing? Well, of course a theatre/cinema needed 'proper' curtains. Curtains that could be remotely opened and closed from the 'projection booth'. There would need to be spotlights and footlights. Romantic music, such as Sleepy Lagoon (remembered from my schoolboy afternoons in the Ritz cinema, Newtownards), would need to be played before the film started. So, curtains first. Ann had been a Domestic Science teacher. Ever resourceful, as well as long-suffering, she remembered a bundle of rich-red curtain material that reposed in the loft, and soon 'ran up' a made-to-measure pair.

A corded curtain-track was purchased after experiments with a home-made one were abandoned due to it having excessive friction. That font of all knowledge, Google, was consulted for prices of motorised curtain units. They were all considered to be too expensive; there are limits to long-sufferance! Therefore the 'model engineering hat' was donned. Very low revolutions per minute were needed, thus worm and pinion gearing of a low voltage DC motor was the way to go. A short length of ½" diameter Whitworth-threaded screwed rod was bored to a close fit on the motor shaft and retained with a grub screw.

Plate 42. Worm and pinion gearing

Plate 43. Curtain-cord tensioning device

With a ½" Whitworth tap gripped in the slowly rotating ML7 three-jaw chuck, a brass disc mounted on a pin on the cross-slide was fed in against the tap, until a satisfactory worm wheel was achieved. Suitable brass side plates were made and arranged such that the correct depth of engagement was maintained between the worm and pinion. This mechanism was mounted on a short ¾" ply swinging-arm so that its weight was supported by a loop of the curtain cord passing round a grooved plywood wheel. See Plates 42 and 43. After some experimentation with the diameter of the wheel and additional weight, correct operation was achieved. When the curtains are either open or shut the cord slips on the wheel without overloading the motor.

A number, I think it was five, 12 volt PAR16 mini-spotlights known as 'Birdies', were purchased from a stage lighting supplier on the internet. Although only two, fitted with red gels, were needed for the stage lighting; they were so attractive that the remainder were given away as Christmas presents to my sons and daughters. A stereo amplifier and speakers (placed on either side of the screen) were purchased at a bargain price from Maplin Electronics. The amplifier input is fed from a record-turntable or a DVD player via an audio changeover switch. To add to the ambiance, a string of programmable multi-coloured low-voltage Christmas tree lights were draped around the periphery of the room. Near the entrance door a mini mirror-ball was hung to provide a swirl of blue, red and green light on surrounding surfaces. Above the door an illuminated 'EXIT' sign was mounted. I had made this many years ago and it was a left-over part from an architectural metal project. The sign was created by removing the letter shapes from a sheet of aluminium using a scroll saw. This was done because the letters had to be in Times New Roman font!

These various efforts were rewarded; on

passing through the entrance door the closed red curtains illuminated by red spots, the flecks of colour from the mirror-ball and the twinkling string of lights assail the eyes. From behind the curtains come the strains of the 'Warsaw Concerto'. See Plate 44. Soon the music fades and the spotlights dim as the curtains part. Pointing fingers of the Twentieth-Century Fox searchlight logo sweep the screen accompanied by the triumphal audio signature.

Plate 44. Grandpa's theatre

An invitation was sent to the family to visit Grandpa's cinema at a secret location one evening just before Christmas 2010. Great excitement was generated amongst the grandchildren by the idea of a 'secret location', especially when they heard they were to be taken there by train. The plan was that the visitors would enter by the front door of the bungalow. As soon as they were all inside, the garage door would be opened and the connecting track quickly laid. The first group of passengers would then exit through the kitchen door at the rear of the house and board the festive train. Travelling anti-clockwise around the loop a 'distant' church and house can be seen on the right glowing in the darkness from internal tea-lights. Over the click of the train wheels can be heard the strains of a familiar Christmas carol.

After one circuit, the train comes to a halt and the jovial driver, who is not Santa Claus despite the red hat with a white bauble, invites the passengers to detrain and visit the shed beside the track. Inside is a 'Snow Village' display. Nestling amongst festoons of holly and ivy and set in deep cotton-wool snow is a model village of lighted porcelain houses. There are cars in the street, people and Christmas trees. Having gazed in wonder, the passengers return to the waiting train.

With an expressive 'ding-ding' of the tram gong the travellers enjoy a spirited run in the darkness along the side of the house. After a stop at the roadway and a reset of the points, the train comes to a final halt in the garage. Carefully avoiding the projecting garage door the passengers ascend the stairs to 'Grandpa's cinema'. Once through the 'magic portal', that is the fire door, refreshments are available from 'Uncle Bob's Bar', named after a much-missed younger brother. 'Bushmills Bluey' with its coaches has already departed to pick up another set of passengers at the back of the house. Once all are seated this evening's show begins, a classic episode

The Darling Buds of May TV series, appropriately entitled 'Christmas is Coming'.

After a dreary winter, it was good to think about the coming of longer days. Santa artwork was replaced by Easter motifs on the 'Celebration Coach' and bookings were taken for family birthdays. These were often for train trips combined with cinema showings of favourite movies. Sometimes, alone, on a summers evening, 'Bluey' would be brought out of the shed and a quiet pint or two of home-brewed beer enjoyed as we rumbled around between the fronds of garden foliage. Perhaps another set of points inserted just here, might switch us to that other time-track where we could always 'live the dream'?

Appendix I

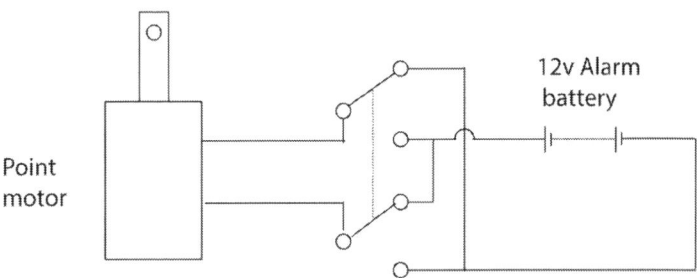

DPDT Point-switch wiring

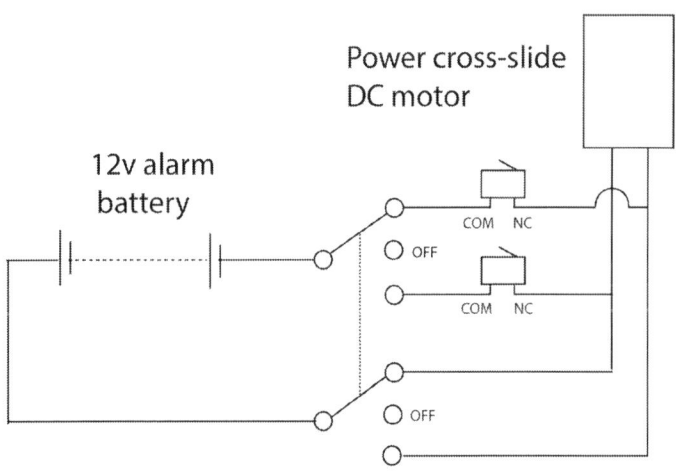

DPDT Switch and limit switch wiring for lathe power cross-slide

Appendix II

```
# Ghost Train Python Listing
from threading import Thread
from time import sleep
import atexit, os, pygame, time
import RPi.GPIO as GPIO

GPIO.setmode(GPIO.BOARD)
GPIO.setup(5, GPIO.IN)
GPIO.setup(11, GPIO.OUT)
GPIO.setup(13, GPIO.OUT)
pygame.mixer.init()

##################
# LOAD TRACKS
##################
trackId = 0
tracks = []
path=r'/home/pi/sounds'
for dir_entry in os.listdir(path):
    trackId=trackId+1
    tracks.append(dir_entry)
    print("found Sound: " + dir_entry)
print("found " + `trackId` + " tracks!")

##################
# PLAY TRACK
##################
def playSound(trackNum):
    track = tracks[trackNum]
    print("Playing Sound: " + track)
```

```python
    pygame.mixer.music.load('sounds/' + track)
    pygame.mixer.music.play(0, 0.0)
    while pygame.mixer.music.get_busy():
        pygame.time.Clock().tick(10)

##################
# TURN ON LIGHT
##################
def toggleLight(delay):
    GPIO.output(11, True)
    GPIO.output(13, True)
    time.sleep(delay)
    GPIO.output(11, False)
    GPIO.output(13, False)

##################
# CLEAN UP
##################
def cleanup():
    print('Exit')
    GPIO.cleanup()

##################
# MAIN LOGIC
##################
try:
    delay=3;
    trackNum=0;
    while True:
        if not GPIO.input(5):
            t1 = Thread(
                target=toggleLight, args=(delay,))
            t2 = Thread(
```

```python
                target=playSound, args=(trackNum,))
            t1.start()
            t2.start()
            t1.join()
            t2.join()
            if trackNum < trackId-1:
                trackNum = trackNum+1
            else :
                trackNum = 0
                sleep(0.1)
except KeyboardInterrupt:
    print('\nExiting...')

atexit.register(cleanup)
```

Abbreviations

amp = Ampere

BA = British Association

BSF = British Standard Fine

CE = European Conformity

CNC = Computer numerical controlled

DC = Direct current

DPDT = Double pole double throw (electric switch)

hp = Horsepower

ID = Internal diameter

LGB = Lehmann Gross Bahn

mA = Milliamp

mph = Miles per hour

OD = Outside diameter

PAR = Planed all round

RCD = Residual current device

rpm = Revolutions per minute

SWG = Standard Wire Gauge

Printed in Great Britain
by Amazon